S. Manigandan
Suresh Norman

Fall Detection System For Elderly Person Monitoring Using GSM Network

S. Manigandan
Suresh Norman

Fall Detection System For Elderly Person Monitoring Using GSM Network

LAP LAMBERT Academic Publishing

Impressum / Imprint
Bibliografische Information der Deutschen Nationalbibliothek: Die Deutsche Nationalbibliothek verzeichnet diese Publikation in der Deutschen Nationalbibliografie; detaillierte bibliografische Daten sind im Internet über http://dnb.d-nb.de abrufbar.
Alle in diesem Buch genannten Marken und Produktnamen unterliegen warenzeichen-, marken- oder patentrechtlichem Schutz bzw. sind Warenzeichen oder eingetragene Warenzeichen der jeweiligen Inhaber. Die Wiedergabe von Marken, Produktnamen, Gebrauchsnamen, Handelsnamen, Warenbezeichnungen u.s.w. in diesem Werk berechtigt auch ohne besondere Kennzeichnung nicht zu der Annahme, dass solche Namen im Sinne der Warenzeichen- und Markenschutzgesetzgebung als frei zu betrachten wären und daher von jedermann benutzt werden dürften.

Bibliographic information published by the Deutsche Nationalbibliothek: The Deutsche Nationalbibliothek lists this publication in the Deutsche Nationalbibliografie; detailed bibliographic data are available in the Internet at http://dnb.d-nb.de.
Any brand names and product names mentioned in this book are subject to trademark, brand or patent protection and are trademarks or registered trademarks of their respective holders. The use of brand names, product names, common names, trade names, product descriptions etc. even without a particular marking in this work is in no way to be construed to mean that such names may be regarded as unrestricted in respect of trademark and brand protection legislation and could thus be used by anyone.

Coverbild / Cover image: www.ingimage.com

Verlag / Publisher:
LAP LAMBERT Academic Publishing
ist ein Imprint der / is a trademark of
OmniScriptum GmbH & Co. KG
Heinrich-Böcking-Str. 6-8, 66121 Saarbrücken, Deutschland / Germany
Email: info@lap-publishing.com

Herstellung: siehe letzte Seite /
Printed at: see last page
ISBN: 978-3-659-76465-3

Zugl. / Approved by: Chennai,Anna University,Diss.,2015

Copyright © 2015 OmniScriptum GmbH & Co. KG
Alle Rechte vorbehalten. / All rights reserved. Saarbrücken 2015

FALL DETECTION SYSTEM FOR ELDERLY PERSONS USING GSM NETWORK

S. Manigandan[1], Suresh R. Norman[2]

mani.sivkar@gmail.com[1], sureshrnorman@ssn.edu.in[2]

ACKNOWLEDGEMENT

I thank my parents for the support and the love whose positive attitude towards education has been an important driving force for my studies.

TABLE OF CONTENTS

CHAPTER NO.	TITLE	PAGE NO.
	LIST OF TABLES	04
	LIST OF FIGURES	05
	LIST OF ABBREVIATIONS	07
1	INTRODUCTION	08
	1.1 MOTIVATION	08
	1.2 ORGANISATION OF THE REPORT	09
2	LITERATURE SURVEY	10
	2.1 LITERATURE REVIEW	10
	2.2 PROPOSED METHOD	12

3	**HARDWARE DESCRIPTION**	14
	3.1 ARDUINO UNO BOARD	14
	3.2 ATMEGA 328 MICROCONTROLLER	15
	3.2.1 Atmega 328 ports	16
	3.2.2 Analog to Digital convertor	17
	3.3 GYRO SENSOR	17
	3.3.1 MPU 6050 operation	19
	3.4 MICROPHONE SENSOR	21
	3.4.1 KY-308 Working principle	22
	3.5 PIEZO VIBRATION SENSOR	23
	3.5.1 Working of piezo sensors	24
	3.5.2 Designing with Piezo sensors	26
	3.6 RF TRANSMITTER	29
	3.7 RF RECEIVER	30
	3.8 GSM MODULE	32
	3.8.1 Block Diagram of GSM Module	32
	3.8.2 AT Commands	33
	3.8.3 Types of AT Commands	35
	3.8.4 Sending the Message	35
	3.9 ENCODER	36
	3.10 DECODER	38
	3.11 POWER SUPPLY	39
4	**SOFTWARE DESCRIPTION**	41
	4.1 WRITING SKETCHES	41
	4.2 ARDUINO IDE COMMANDS	41
	4.3 SKETCHBOOK	43
	4.4 UPLOADING	44

	4.5 LIBRARIES	45
	4.6 SERIAL MONITOR	46
5	**SYSTEM IMPLEMENTATION**	47
	5.1 TRANSMITTER SECTION	47
	5.2 RECEIVER SECTION	49
	5.3 FLOWCHART OF THE SYSTEM	50
	5.4 RESULT AND DISCUSSIONS	51
	5.5 FALL DETECTION IN ADL	55
6	**CONCLUSION AND FUTURE WORK**	57
	6.1 CONCLUSION	57
	6.2 FUTURE ENHANCEMENT	57
	APPENDIX	58
	REFERENCES	59

LIST OF TABLES

TABLE NO	TITLE	PAGE NO
3.1	Arduino Uno Board Specifications	14
3.2	MPU-6050 Absolute Maximum Ratings	20
3.3	KY-308 Sensor Specifications	21
3.4	Piezo Vibration Sensor Specifications	24
3.5	RF 433MHz Transmitter Specifications	30
3.6	RF 433MHz Receiver Specifications	31
3.7	Encoder HT12E Specifications	37
3.8	Decoder HT12D Specifications	39
5.1	Analysis of prototype testing	52
5.2	Accuracy, Specificity, Sensitivity of the falls	52
5.3	Threshold levels for Sensors	53
5.4	Fall detection	55

LIST OF FIGURES

FIG NO	TITLE	PAGE NO
3.1	Pin Diagram of Atmega 328 Microcontroller	16
3.2	System Diagram of MPU 6050	18
3.3	Pin Diagram of MPU 6050	20
3.4	Pressure applied on constrained mass	25
3.5	Pressure applied on unconstrained mass	26
3.6	Signal conditioning circuit of piezo sensor	27
3.7	Piezo vibration sensor	28
3.8	Sensitivity of vibration sensor	28
3.9	Pin Diagram of RF 433MHz Transmitter	29
3.10	Pin Diagram of RF 433MHz Receiver	31
3.11	Block Diagram of GSM SIM900 Module	33
3.12	Pin Diagram of Encoder	37
3.13	Pin Diagram of Decoder	38
4.1	Arduino Sketchbook	43
4.2	Importing a Library into Arduino IDE	45
4.3	Serial Monitor	46

5.1	Output of MPU 6050	47
5.2	Output of Piezo vibration sensor	48
5.3	Block Diagram of Transmitter Section	49
5.4	Block Diagram of Receiver Section	49
5.5	Flowchart of the System	50
5.6	Output for test condition	54

LIST OF ABBREVIATIONS

ABBREVIATIONS	EXPLANATION
SMS	Short Message Service
IR	Infra Red
GSM	Global System for Mobile communications
RF	Radio Frequency
ADC	Analog to Digital Converter
IDE	Integrated Development Environment
USB	Universal Serial Bus
SRAM	Static Random Access Memory
FSK	Frequency Shift Keying
ASK	Amplitude Shift Keying
VCO	Voltage Controlled Oscillator
PLL	Phase Locked Loop
SIM	Subscriber Identity Module
TTL	Transistor-Transistor Logic
HTML	Hyper Text Markup Language
GPRS	General Packet Radio Service
UDP	User Datagram Protocol
DDS	Direct Digital Synthesizer
LED	Light Emitting Diode
MPU	Motion Processing Unit
DOF	Degree Of Freedom

CHAPTER 1

INTRODUCTION

Falls, occurring in the elderly are a major public concern to our modern societies. Elderly human beings falling to the ground require immediate medical attentions as one tenth of such falls result in fractures. When they fall, they can become disoriented, immobilized, or knocked unconscious and unable to call for help. Thus fall detection is a critical event, requiring quick and accurate response especially for elderly people living by themselves. Throughout the world there is an increasing need for ambulatory monitoring of physiology to enable medical interventions and therapies outside of the clinical setting. The objective is to employ an assistive device for elderly persons to monitor their health and alert about any emergency conditions if it, arises to the caregiver through SMS.

1.1 MOTIVATION

The main reason that pushed for the development of the presented device is to allow older people to live safely in their own houses as long as possible. This is important not only for health aspects regarding the assisted people, but also for the consequent social advantages such as the possibility for care-giver institutions to employ more efficient and optimized services at lower cost and hence available to a wider part of the population. This will help the children of elderly to monitor physiological condition of parents from their working site. Device which gives the position of a person like accelerometers are available widely and hence deployed in most of the fall detection systems. Automatic devices to classify between the daily activities and emergency situation seem to be a good solution to reach this objective.

In particular, recent miniaturization, cost reduction of MEMS sensors and availability of reliable wireless communication technologies enabled affordable wearable monitoring systems that can be worn by people performing their normal daily activities. For these reasons, in the last few years, the use of portable devices in the health monitoring of chronic patients has considerably increased. Furthermore, in order to prevent false or missed alarms, the use of multi-sensor systems is reliable and the wearable sensors have the fundamental role to validate system decisions.

1.2 ORGANISATION OF THE REPORT

The following gives an overview about every chapter that is being discussed in this report. The chapters are organized as follows

- Chapter 2 takes into account the works done previously by several authors. It gives details about the relative devices and suggestions made from time to time. In this chapter a description about the proposed work is also referred.
- In Chapter 3, the hardware components that are used in this project, their working and specifications to be used are given.
- Chapter 4 gives details about the software used and its usage in this project.
- Ideally, chapter 5 describes the implementation of the proposed work, experimental results and discussions.
- Finally, chapter 6 provides the pros and cons of previous works. It also includes the future enhancements to be made in this project.

CHAPTER 2

LITERATURE SURVEY

2.1 LITERATURE REVIEW

Norbert Noury et al., (2000) [1] used pyro-electric infrared sensor and a magnetic contact to detect the approach or the passage of the person. The IR detector (DP8111) of the thermal type detects the displacements of the person. The magnetic contact switches (DP8211), installed on some doors, can detect its opening and shutting. After a level triggering, a Boolean data is generated for "position", "fall" and "vibrations" which is further interpreted to determine the persons situation. The system is implemented with Atmel processor (AT9OS8535). The communication with the local personal computer is performed via a RF transceiver.

Gerard M. Lyons et al., (2006) [2] developed a system to measure the mobility by an accelerometer based portable unit, worn by each monitored subject. Mobility level summaries are transmitted hourly, as an SMS message, directly from the portable unit to a remote server for long-term analysis. The portable unit houses the Analog Devices ADuC812S microcontroller board, a GSM modem, and a battery-based power supply. Two integrated accelerometers (ADXL202) are connected to the portable unit through the analog inputs of the microcontroller. Each subject's mobility levels are monitored using custom-designed mobility alert software, and the appropriate medical personnel are alerted by SMS if the subject's mobility levels decrease. The program used for analysis is developed using C++ and a relational database which is flexible is created using SQL.

Tong Zhang et al., (2006) [3] in their experiment used a system that consists of two computers and a cell phone with a box affixed to it. The computers are used as remote servers. Inside the box, the circuit includes a tri-axial accelerometer (Freescale MMA7260Q), a single chip modem (OKI semiconductor's MSM7512BRS), a microcontroller unit (Microchip's PIC18F2455) and some other peripheral components. The circuit and the cell phone compose a fall detector: the acceleration signals are transformed into digital signals via A/D conversion, and then analysed by the MCU. If a fall is determined, the MCU will send a UDP package to a remote server via the wireless channel. Here, server 1 is the master server, and server 2 is the reserve server. And on the other hand, the MCU can also receive the instructions and data from either of the servers. The software includes the MCU program, the server program and a database.

N.Noury et al., (2007) [4] have dealt at length fall detection principles and methods based on analytical methods and machine learning methods. This paper points out the difficulty to compare the performance of the different systems due to the lack of a common framework. They have classified the fall events into various categories such as true positive, false positive, true negative, false negative. In their work they used floor sensitive sensors, vibration sensors, IR sensor, stereovision techniques for analytical analysis. Machine learning methods used some simulated sequences, a neural network can be deployed to discriminate the fall events.

Fletcher et al., (2010) [5] carried out their work using wearable sensors. Wearable sensors have been successfully integrated into clothing garments as well as fashion accessories such as hats, wrist bands, socks, shoes, eyeglasses and headphones. This enables long term continuous monitoring of physiological condition. These systems often include temperature sensors and accelerometers, which are often used to monitor and classify a person's physical activity. In addition to these, the authors has explored sensors that are fundamental to psychophysiology and understanding of human emotions.

J.Wang et al., (2014) [6] proposed an improved fall detection system whose core structure is based on a MCU. The accelerometer sensor is complemented by other smart sensors including temperature and humidity sensors all integrated on one single board, recording real time acceleration and ambient environment information. Both acceleration and environment information are first captured using an ADC. Then, the digital signal is transmitted to the MCU for further processing. The heart rate is captured by a pulse pressure sensor and also passed directly to the MCU. The system is complemented with a customer interface designed to monitor information in real-time.

Quoc T. Huynh et al., [7] developed a wireless sensor system and an algorithm to identify the fall events compared to normal ADLs. The system includes a Wireless Sensor System (WSS) and a detection algorithm. The WSS transmits and receives real-time accelerometer and gyro data during the fall. The detection algorithm is based on a simple threshold method. The wireless sensors system contains a set of Sensor module, Micro Control Unit, and Wi-Fi module used to sense body orientation and activity data, control the flow of data, and transmit/receive data, respectively. The WSS is placed at the center of chest.

2.2 PROPOSED METHOD

The fall detection system as suggested by J.Wang et al., (2014) has been effective when we combine and use two or more conventional approaches. This reduces the amount of false alarms and increases the efficiency to 90% - 95%. The proposed method has been tested using a prototype system.

Choosing which sensors to be employed and determining the threshold level of the test conditions to be measured using the sensors decides the scope of the entire systems functioning. To detect the position and orientation of a person, gyro sensors are most preferred. They are compact as well as low cost. The accelerometer readings alone are not sufficient to determine the fall detection of the elderly. Because it show steep change in the values when a person sits suddenly.

The choice of a support system to detect the fall event using passive sensors avoids most false alarms. Microphone sound sensors are employed for this purpose in this project as they are not affected by thermal interferences. Moreover the impact of the fall can be sensed reliably by using a piezo vibration sensor.

A pair of 433MHz RF transmitter and receiver proves to be a cost effective wireless medium to be employed. Upon reaching the test conditions, a GSM module interfaced to the receiver section will send an emergency alert to the care-giver. The transmitter section consists of a microcontroller to sense the sensor values and RF transmitter to transmit these values to the receiver section. The receiving section consists of a microcontroller board to read the RF receiver values and a GSM module to send alert to the care-giver upon reaching the threshold values set for emergency alert.

The program for testing the system modules and entire system is written in Embedded C. Using Embedded C programs saves the memory space in microcontroller. The codes are compiled, verified and uploaded to microcontrollers using the Arduino IDE. For this purpose a USB cable (Type A to Type B) is needed for the Arduino Uno board. The power supply to arduino boards and GSM module can be provided by means of 12V DC adapter for stable operation of the devices.

CHAPTER 3

HARDWARE DESCRIPTION

3.1 ARDUINO UNO BOARD

The Arduino Uno is a microcontroller board based on the Atmega328. It has a 16 MHz ceramic resonator, a USB connection, a power jack, an ICSP header, and a reset button. It contains everything needed to support the microcontroller; simply connect it to a computer with a USB cable or power it with an AC-to-DC adapter or battery to get started. The Uno differs from all preceding boards as it does not use the FTDI USB-to-serial driver chip. Instead, it features the Atmega16U2 programmed as a USB-to-serial converter.

Table 3.1 Arduino Uno Board Specifications

UNO HARDWARE SPECIFICATIONS	
Microcontroller	Atmega328
Digital I/O Pins	14
Analog Input Pins	6
Flash Memory	32KB
SRAM	2KB
EEPROM	1KB

Revision 3 of the Arduino Uno board has the following new features
- Stronger RESET circuit.
- Atmega 16U2 replace the 8U2.
- pin out: added Serial Data and Serial Clock pins that are near to the AREF pin and two other new pins placed near to the RESET pin, the IOREF that allow the shields to adapt to the voltage provided from the board.

3.2 ATMEGA328 MICROCONTROLLER

The high performance Atmel 8-bit AVR RISC-based microcontroller combines 1KB EEPROM, 2KB SRAM, internal and external interrupts, serial programmable USART, The device operates between 1.8-5.5 volts. The Atmega328P is supported with a full suite of program and system development tools including: C Compilers, Macro Assemblers, Program Debugger/Simulators, In-Circuit Emulators, and Evaluation kit.

By executing powerful instructions in a single clock cycle, the device achieves throughputs approaching 1 MIPS per MHz, balancing power consumption and processing speed. The salient features of Atmega328 are
- Four ports with 23 programmable I/O lines.
- Advanced RISC architecture.
- High endurance non-volatile memory segments.
- 32KB ISP flash memory with read-while-write capabilities.
- 32 general purpose working registers.
- A byte-oriented 2-wire serial interface.
- Six channel 10-bit ADC programmable.
- Six PWM channels.

- Programmable serial USART.
- Watch Dog Timer with internal oscillator.

3.2.1 Atmega 328 Ports

Atmega 328 has four ports in it. Port B is an 8-bit bi-directional I/O port with internal pull-up resistors. Depending on the clock selection fuse settings, PB6 can be used as input to the inverting oscillator amplifier and input to the internal clock operating circuit. Depending on the clock selection fuse settings, PB7 can be used as output from the inverting oscillator amplifier. Port C is a 7-bit bi-directional I/O port with internal pull-up resistors. Port D is an 8-bit bi-directional I/O port with internal pull-up resistors. The output buffers of these ports have symmetrical drive characteristics with both high sink and source capability. As inputs, port pins that are externally pulled low will source current if the pull-up resistors are activated. The port pins are tri-stated when a reset condition becomes active, even if the clock is not running. Fig 3.1 shows various port pins in Atmega328 microcontroller.

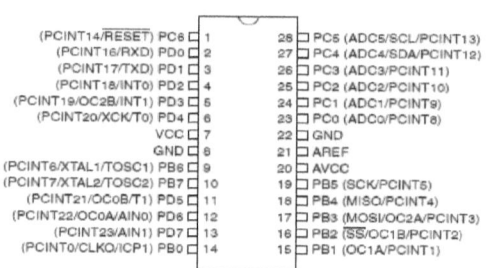

Fig 3.1 Pin Diagram of Atmega328 Microcontroller

3.2.2 Analog to Digital Converter

The Atmega328P features a 10-bit successive approximation ADC. The ADC is connected to an 8-channel analog multiplexer which allows eight single-ended voltage inputs constructed from the pins of Port A.

The single-ended voltage inputs refer to 0V (GND). The ADC contains a sample and hold circuit which ensures that the input voltage to the ADC is held at a constant level during conversion. The ADC has a separate analog supply voltage pin, AVCC. AVCC must not differ more than ±0.3V from VCC. Internal reference voltages of nominally 1.1V or AVCC are provided. The voltage reference may be externally decoupled at the AREF pin by a capacitor for better noise performance.

The Power Reduction ADC bit must be disabled by writing a logical zero to enable the ADC. The ADC converts an analog input voltage to a 10-bit digital value through successive approximation. The minimum value represents GND and the maximum value represents the voltage on the AREF pin minus 1 LSB. Optionally, AVCC or an internal 1.1V reference voltage may be connected to the AREF pin by writing to the REFSn bits in the ADMUX Register. The internal voltage reference may thus be decoupled by an external capacitor at the AREF pin to improve noise immunity.

3.3 GYRO SENSOR

Gyroscopes measure the angular rate of rotation about one or more axes. Gyroscopes can measure complex motions accurately in free space, hence, making it a required motion sensor for tracking the position and rotation of a moving object. Unlike accelerometers and compasses, gyroscopes are not dependent on any external forces such as gravity or magnetic fields, and can therefore function fairly autonomously.

In this project the gyro sensor used is MPU-6050 manufactured by Invensense incorporation. The MPU-6050 is a 6-axis motion tracking devices designed for the low power, low cost, and high performance requirements of smart-phones, tablets and wearable sensors.

The MPU-6050 incorporates motion fusion and run-time calibration firmware that enables to eliminate the costly and complex selection, qualification, and system level integration of discrete devices in motion-enabled products, and guarantees that sensor fusion algorithms and calibration procedures deliver optimal performance for consumers.

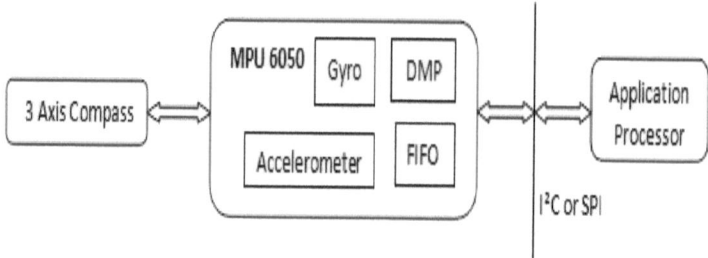

Fig 3.2 System Diagram Of MPU-6050

The MPU-6050 devices combine a 3-axis gyroscope and a 3-axis accelerometer on the same silicon die together with an onboard Digital Motion Processor (DMP) which processes complex 6-axis motion fusion algorithms. The device can access external magnetometers or other sensors through an auxiliary master I²C bus, allowing the devices to gather a full set of sensor data without intervention from the system processor. The devices are offered in the same 4x4x0.9 mm QFN footprint and pin-out, providing a simple upgrade path and making it easy to fit on space constrained boards. For precision tracking of both fast and slow motions, the parts feature a user-programmable gyro full-scale range of ±250, ±500, ±1000, and ±2000°/sec (dps) and a user-programmable accelerometer full-scale range of ±2g, ±4g, ±8g, and ±16g.

3.3.1 MPU-6050 Operation

The gyroscope sensor is a solid state device based on the Coriolis effect to measure angular rate. The sensor consists of a micro-machined vibrating silicon element, which provides an analogue output proportional to the rate of change of angular position about its sensing axis.

The silicon element design is based on the principle of a shell resonator whereby a silicon ring is mounted to a substrate by radial spokes. This design of resonator is more resilient to shock and vibration than beam oscillators whereby the suspended silicon beam is mounted vertically with legs supporting the element mass. The Gyros are available to measure a single axis or three axes.

The gyroscope is not free from noise, however, because it measures, rotation it is less sensitive to linear mechanical movements. The type of noise that accelerometer suffers from, however gyroscopes have other types of problems like for example drift (not coming back to zero rate value when rotation stops). Nevertheless by averaging data that comes from accelerometer and gyroscope we can obtain a relatively better estimate of current device inclination than we would obtain by using the accelerometer data alone.

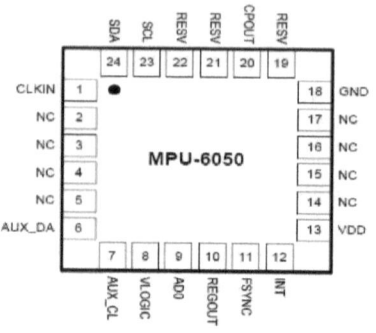

Fig 3.3 Pin Diagram of MPU-6050

The gyroscope data are filtered using kalman filters to get averaged value. Now much simpler complimentary filters are used. The absolute maximum ratings of gyro sensor are specified in Table 3.2.

Table 3.2 MPU-6050 Absolute Maximum Ratings

MPU-6050 ABSOLUTE MAXIMUM RATINGS	
Acceleration(Any axis, Unpowered)	10,000g for 0.2ms
REGOUT	-0.5V to 2V
Supply Voltage	-0.5V to 6V
CPOUT(2.5V<Vdd<3.6V)	-0.5V to 30V
Temperature Range(Powered)	-40°C to +105°C
Temperature Range(Storage)	-40°C to +125°C
Sensitivity	270 mV/g to 330 mV/g

3.4 MICROPHONE SOUND SENSOR

This sound sensor detector is small board that combines a microphone and some processing circuitry. Most sensors only provide the analog value, so it outputs a voltage value to indicate the sensing parameters. Arduino reads this value from any analog pin A0 to A5, ranging from 0 till 1023. In AVR, the analog voltage varies from 0V till 5V. We sign AO (Analog Output) as a pin name on this project. Sometimes we only want the sensors only give feedback when the sensing value read a threshold that we want, so when it reached the feedback is 1, and 0 vice verse. This can be observed using onboard LED in arduino.

The microphone sound sensor consists of three stages. The first section of the circuit is the microphone capsule. The second stage of the circuit is an envelope follower. The final stage implements a thresholded switch (such as Schmitt trigger) on the envelope signal. The output of the Schmitt trigger is found on the gate pin can be connected to a digital input which can be used to trigger interrupts.

Table 3.3 KY-308 Sensor Specifications

KY-3O8 SENSOR SPECIFICATIONS	
Working Voltage	4-6V(DC)
Working Current	15mA(maximum)
Working Frequency	40Hz
Output Signal	0-5V
High-Accuracy	0.3cm
Echo Signal	TTL PWM Signal

3.4.1 KY-308 working principle

The heart of the sound detector is the microphone capsule which converts acoustic energy into electrical energy. Inside the capsule is the diaphragm, which is actually one plate of a small capacitor. That capacitor forms a voltage divider with the external bias resistor. The diaphragm moves in response to sound, and the capacitance changes as the plates get closer together or farther apart, causing the divider to change. Since capacitors are sensitive to loading, it's internally buffered with a JFET. Due to the mechanical and electronic tolerances involved, some capsules are more sensitive than others. Also, the JFET is rather sensitive to noise on the power supply. Both of these factors need to be accounted for when deploying the sound detector.

The sound detector is an analog circuit, and as such, it's more sensitive to noise on the power supply than most digital circuits. Since the capsule is effectively a voltage divider sitting across the power rails, it will transcribe any noise on the supply lines onto the capsule output. The next circuit in the chain is a high-gain amplifier, so any noise on the supply will then be amplified. Therefore, the sound detector may require more careful power supply configuration than many circuits.

In testing with various supplies, a significant degree of variability was discovered - some supplies are less noisy than others. When ripples are as high as 30 mV on the supply output, the sound detector was rather sensitive and unstable. We can check how clean a power supply is by checking it with an oscilloscope or volt meter, set to the AC Volts (or, if provided, the AC millivolts) range. A truly clean supply will show 0.000 VAC. Based on the supplies used in testing, ripple of more than about 10 mV is problematic.

Powering arduino with a 9V external supply, which allows the onboard regulators to function, the Arduino's 5V output was sufficiently clean. However, powering it from the 5V available on the USB port on a PC, the regulators are bypassed, and the results were somewhat less usable, and vary greatly between different ports on different PCs.

A powered USB hub will probably provide cleaner power than the ports on the PC itself. If all else fails, three 1.5V batteries in series make a nice, clean source of 4.5V.

The sound detector comes set for moderate sensitivity - speaking directly into the microphone, or clapping your hands nearby should cause the gate output to fire. The gain is set by changing the feedback resistors in the preamp stage.

3.5 PIEZOVIBRATION SENSOR

The piezoelectric sensor is used for flex, touch, vibration and shock measurement. Its basic principal, at the risk of oversimplification, is as follows: whenever a structure moves, it experiences acceleration.

A piezoelectric shock sensor, in turn, can generate a charge when physically accelerated. This combination of properties is then used to modify response or reduce noise and vibration.

Most electronic applications use quartz since its growth technology is far along, thanks to development of the reverse application of the piezoelectric effect; the quartz oscillator.

Table 3.4 Piezo Vibration Sensor Specifications

PIEZO VIBRATION SENSOR SPECIFICATIONS	
Voltage sensitivity (open-circuit, baseline)	1.1V/g
Voltage sensitivity (open-circuit, resonance)	6V/g
Charge sensitivity	260pC/g
Resonance frequency	75Hz
Upper limiting frequency(+3dB)	42Hz
Capacitance	244pF
Dissipation factor	0.018
Linearity	+/-1 %
Linear mass	0.3gram
Storage temperature	-40 to 80°C
Operating temperature	-20 to 60°C

3.5.1 Working of Piezo sensors

Minerals such as tourmaline and quartz could transform mechanical energy into an electrical output. The voltage induced from pressure (Greek: piezo) is proportional to that applied pressure, and piezoelectric devices can be used to detect single-pressure events as well as repetitive events.

The ability of certain crystals to exhibit electrical charges under mechanical loading was of no practical use until very-high-input impedance amplifiers enabled to amplify the signals produced by these crystals. Sensors based on the piezoelectric effect can operate from transverse, longitudinal, or shear forces, and are insensitive to electric fields and electromagnetic radiation. The response is also very linear over wide temperature ranges, making it an ideal sensor for rugged environments.

The physical design of the piezoelectric sensor depends on the type of sensor. For example, the configuration of a pressure sensor, or a shock (impulse) sensor, would arrange a smaller, but well-known mass of the crystal in a transverse configuration, with the loading deformation along the longest tracks to a more massive base. This assures that the applied pressure will load the base from only one direction. Figure 3.4 shows a constrained mass is allowed to deform the crystal sensor in one axis. This configuration is good for force and pressure.

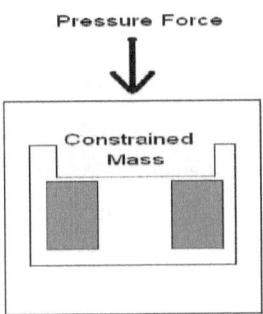

Fig 3.4 Pressure applied on constrained mass

An accelerometer based on the piezoelectric effect, would use a known mass to deform the sensing crystal part in either a positive or negative direction depending on the excitation force. It should be noted that we need to known modulus of elasticity in the sensor substrate. Figure 3.5 shows because the modulus of elasticity is known for a substrate material, the unconstrained mass is allowed to move with vibration making this type of piezoelectric sensor ideal for detecting shock and vibration.

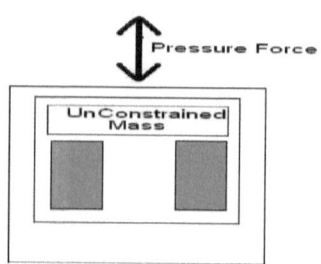

Fig 3.5 Pressure applied on unconstrained mass

3.5.2 Designing with piezo sensors

Piezoelectric sensors require some precautions when connecting to sensitive electronic components. First and foremost, the voltage levels created by hard shock can be-very high even around 100V spikes.

More than likely, an op amp will be used to interface these sensors to an A/D converter, either discrete or on a microcontroller. One tip is to choose a high-input-impedance op amp to minimize current. One possible candidate is the linear technology JFET input dual op amp. It has 10^{12} Ω input resistance and a 1 MHz gain bandwidth product, good enough to easily handle the vibration ranges of this sensor.

Another suitable part is the TLV2771 from Texas Instruments. This rail-to-rail low-power op-amp also has a 10^{12} Ω differential input resistance and a 5 MHz unity-gain bandwidth. Signal conditioning in a single stage can prepare the input from the shock sensor directly into an A/D converter (Figure 3.6).

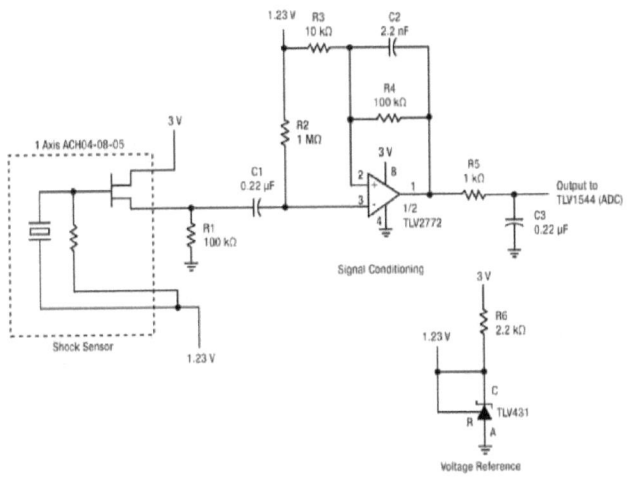

Fig 3.6 Signal conditioning circuit of piezo sensor

Figure 3.6 shows Op amps such as the TI TLV2772 feature high input impedances to help minimize current from the potentially high-voltage inputs from the piezoelectric sensors. Op-amp circuits can be designed to operate in voltage mode or charge mode. Charge mode is used when the amplifier is remote to the sensor. Voltage mode is used when the amplifier is very close to the sensor. Another tip is to attenuate the input signal and use the op amp's gain to bring into the desired range. Snubbing protection on the inputs of the op-amp is needed, especially if the design could be subjected to harsh conditions. This sensor would not only generate a positive voltage, but, in reality, the signal from the sensor can ring and introduce negative voltage spikes We need to squelch negative voltage levels on the op-amp inputs, especially if using only a single rail power supply on the op amp.

Fig 3.7 Piezo Vibration sensor

Adding mass to a piezoelectric sensor can change its resonant frequency as well as change its baseline sensitivity. Figure 6 shows loading a sensor with mass changes its resonant frequency and sensitivity.

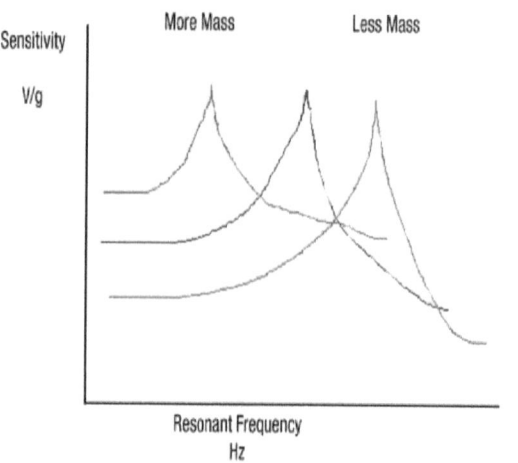

Fig 3.8 Sensitivity of vibration sensor

In addition to sensing vibration and shock, a piezoelectric device can also be used to extract ambient energy. For example the MideV22BL which is a hermetically sealed piezoelectric sensor capable of sensing from 26 to 100Hz vibrations.

3.6 RF TRANSMITTER

The TRF4400 single-chip solution is an integrated circuit intended for use as a low cost FSK transmitter to establish a frequency-agile RF link. The device is available in a 24-lead TSSOP package and is designed to provide a fully-functional multichannel transmitter. The chip is intended for linear (FM) or digital (FSK) modulated applications in the 433-MHz ISM band. The single-chip transmitter operates down to 2.2V and is expressly designed for low power consumption. The synthesizer has a typical channel spacing of approximately 230 Hz to allow narrow-band as well as wide-band applications. Due to the narrow channel spacing of the direct digital synthesizer (DDS), the DDS can be used to adjust the TX frequency and allows the use of inexpensive reference crystals. The transmitter consists of an integrated VCO, a complete fully-programmable direct digital synthesizer, and a power amplifier. The internal VCO can be used with an external tank circuit or an external VCO. The divider, prescaler, and reference oscillator require only the addition of an external crystal and a loop filter to provide a complete DDS with a typical frequency resolution of 230 Hz.

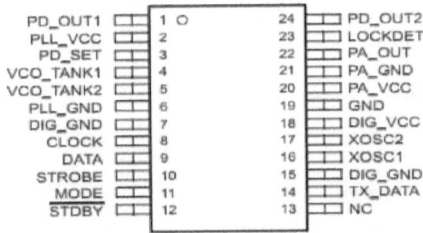

Fig 3.9 Pin Diagram of RF 433MHz Transmitter

The various pins of RF 433MHz transmitter is shown in Fig 3.9. The 8-bit FSK frequency deviation register determines the frequency deviation in FSK mode. The modulation itself is done in DDS so no additional external components are necessary in the circuit. Since the typical RF output power is approximately 7dBm, no external RF power amplifier is necessary in most applications.

The TRF4400 RF transmitter is suitable for use in applications that include the TRF6900 RF transceiver. The operating specifications of RF 433 MHz transmitter are given in Table 3.5.

Table 3.5 RF 433 MHz Transmitter Specifications

RF TRANSMITTER SPECIFICATIONS	
Working Voltage	3V to 12V DC
Working Current	9mA to 40mA
Working Frequency	315MHz or 433MHz
Resonance mode	SAW
Modulation mode	ASK
Transmission Power	25mW(at 12V)

3.6 RF RECEIVER

This is a PLL based ASK Hybrid 433MHz RF receiver module and is ideal for short-range wireless control applications where quality is a primary concern. The receiver module requires no external RF components except for antenna. The easier method to incorporate is to use a simple wire of length about 23cm for 315MHz and 17cm for 434MHz. The super regenerative design exhibits exceptional sensitivity at a very low cost. These wireless receivers work with our 434MHz transmitters. They can easily fit into a breadboard and work well with microcontrollers to create a very simple wireless data link. As these are only receivers, they will only work communicating data one-way, we would need two pairs to act as a transmitter/receiver pair.

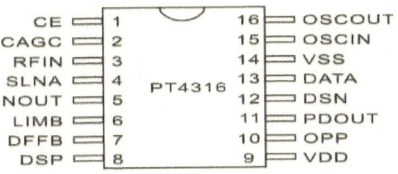

Fig 3.10 Pin diagram of RF 433 MHz Receiver

These modules are indiscriminate and will receive a fair amount of noise. Both the transmitter and receiver work at common frequencies and don't have IDs. Hence, a method of filtering noise and pairing transmitter and receiver is needed.

Table 3.6 RF 433 MHz Receiver Specifications

RF RECEVIER SPECIFICATIONS	
Working Voltage	0.5V to 5V DC
Working Current	less than 5.5 mA
Working Frequency	315MHz or 433.92MHz
Working mode	OOK/ASK
Bandwidth	2MHz
Transmission Power	25mW(at 12V)

3.8 GSM MODULE

SIM900A is an ultra-compact and reliable wireless module-SIM900. This is a complete Quad-band GSM/GPRS module in a SMT type and designed with a very powerful single-chip processor integrating ARM926EJ-S core, allows it to benefit from small dimensions and cost-effective solutions. Featuring an industry-standard interface, the SIM900 delivers GSM/GPRS 850/900/1800/1900MHz performance for voice, SMS, Data, and Fax in a small form factor and with low power consumption. With a tiny configuration of 24mm x 24mm x3 mm, SIM900 can fit almost all the space requirements in M2M applications, especially for slim and compact demands of design.

3.8.1 Block Diagram of GSM Module

This is a GSM Modem with a simple to implement RS232, TTL Serial, Use it to send SMS, make and receive calls, and do other GSM operations by simple AT commands through a serial interface from microcontrollers and computers. It uses the highly popular SIM900A module for all its GSM operations. It comes with a standard RS232 interface which can be used to easily interface the modem to microcontrollers and computers. The modem also features a serial TTL interface option.

All modem operations can be carried out by sending AT commands to virtual serial port though a serial terminal program. Most programming languages allow sending and receiving serial commands to a serial port and can be used to write software that can operate the modem without the need to implement any complex interface. The various functional blocks of GSM module are shown Fig 3.11.

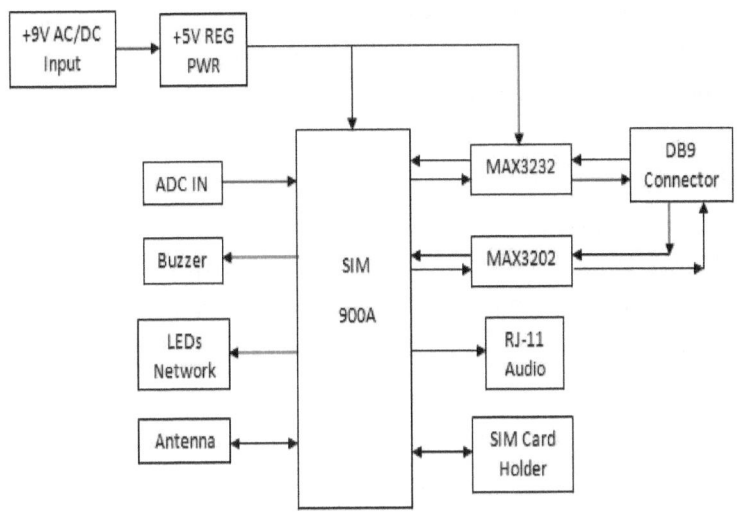

Fig 3.11 Block Diagram of GSM SIM900 Module

3.8.2 AT Commands

AT commands are instructions used to control a modem. AT is the abbreviation of ATtention. Every command line starts with "AT" or "at". Many of the commands that are used to control wired dial-up modems, such as ATD (Dial), ATA (Answer), ATH (Hook control) and ATO (Return to online data state), are also supported by GSM/GPRS modems and mobile phones. Besides this common AT command set, GSM/GPRS modems and mobile phones support an AT command set that is specific to the GSM technology, which includes SMS-related commands like AT+CMGS (Send SMS message), AT+CMSS (Send SMS message from storage), AT+CMGL (List SMS messages) and AT+CMGR (Read SMS messages).

The starting "AT" is the prefix that informs the modem about the start of a command line. It is not part of the AT command name. For example, D is the actual AT command name in ATD and +CMGS is the actual AT command name in AT+CMGS. Here are some of the tasks that can be done using AT commands with a GSM/GPRS modem or mobile phone:

- Get basic information about the mobile phone or GSM/GPRS modem. For example, name of manufacturer (AT+CGMI), model number (AT+CGMM), IMEI number (International Mobile Equipment Identity) (AT+CGSN) and software version (AT+CGMR).
- Get basic information about the subscriber. For example, MSISDN (AT+CNUM) and IMSI number (International Mobile Subscriber Identity) (AT+CIMI).
- Get the current status of the mobile phone or GSM/GPRS modem. For example, mobile phone activity status (AT+CPAS), mobile network registration status (AT+CREG), radio signal strength (AT+CSQ), battery charge level and battery charging status (AT+CBC).
- Establish a data connection or voice connection to a remote modem (ATD, ATA, etc)
- Send (AT+CMGS, AT+CMSS), read (AT+CMGR, AT+CMGL), write (AT+CMGW) or delete (AT+CMGD) SMS messages and obtain notifications of newly received SMS messages (AT+CNMI).
- Send and receive fax (ATD, ATA, AT+F*).Read (AT+CPBR), write (AT+CPBW) or search (AT+CPBF) phonebook entries.
- Perform security-related tasks, such as opening or closing facility locks (AT+CLCK), checking whether a facility is locked (AT+CLCK) and changing passwords (AT+CPWD). Facility lock examples: SIM lock - a password must be given to the SIM card every time the mobile phone is switched on and PH-SIM lock - a certain SIM card is associated with the

mobile phone. To use other SIM cards with the mobile phone, a password must be entered.

- Control the presentation of result codes / error messages of AT commands. For example, we can control whether to enable certain error messages (AT+CMEE) and whether error messages should be displayed in numeric format or verbose format (AT+CMEE=1 or AT+CMEE=2)
- Get or change the configurations of the mobile phone or GSM/GPRS modem. For example, change the GSM network (AT+COPS), bearer service type (AT+CBST), radio link protocol parameters (AT+CRLP), SMS center address (AT+CSCA) and storage of SMS messages (AT+CPMS).
- Save and restore configurations of the mobile phone or GSM/GPRS modem. For example, save (AT+CSAS) and restore (AT+CRES) settings related to SMS messaging such as the SMS center address.

3.8.3 Types of AT commands

There are two types of AT commands: basic commands and extended commands. Basic commands are AT commands that do not start with "+". For example, D (Dial), A (Answer), H (Hook control) and O (Return to online data state) are basic commands. Extended commands are AT commands that start with "+". All GSM AT commands are extended commands. For example, +CMGS (Send SMS message), +CMSS (Send SMS message from storage), +CMGL (List SMS messages) and +CMGR (Read SMS messages) are extended commands.

3.8.4 Sending the message

To check if the modem supports this text mode, we use the following command: AT+CMGF=1<ENTER>. If the modem responds with "OK" this mode is supported. This mode is only possible to send simple text messages. It is not possible to send multipart, Unicode, data and other types of messages. If the modem contains a SIM card with is secured with a PIN code, we have to use the following command: AT+CPIN="PIN CODE"<ENTER>.

After setting the PIN code, wait some seconds before issuing the next command to give the modem some time to register with the GSM network.

In order to send a SMS, the modem has to be put in SMS text mode first using the following command: AT+CMGF=1<ENTER>. To send the SMS message, type the following command: AT+CMGS="PHONE NUMBER" <ENTER>. The modem will respond with: > .Now type the message text and send the message using the <CTRL>-<Z> key combination: Emergency<CTRL-Z>. After some seconds the modem will respond with the message ID of the message, indicating that the message was sent correctly: + CMGS:62. The message will arrive on the mobile phone shortly.

3.9 ENCODER

The details of various pins of Encoder are shown in Fig 3.7. HT12E Encoder IC will convert the 4 bit parallel data given to pins D0 – D3 to serial data and will be available at DOUT. This output serial data is given to ASK RF Transmitter. Address inputs A0 – A7 can be used to provide data security and can be connected to GND (Logic ZERO) or left open (Logic ONE). HT12E is able to operate in a wide voltage range from 2.4V to 12V and has a built in oscillator which requires only a small external resistor. Its power consumption is very low, standby current is 0.1μA at 5V VDD and has high immunity against noise.

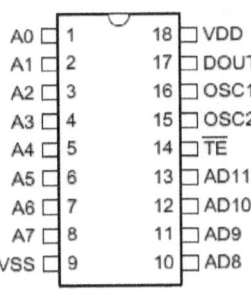

Fig 3.12 Pin Diagram of Encoder

Status of these Address pins should match with status of address pins in the receiver for the transmission of the data. Data will be transmitted only when the Transmit Enable pin (TE) is LOW. 1.1MΩ resistor will provide the necessary external resistance for the operation of the internal oscillator of HT12E. We can provide 8 bit security code for data transmission and multiple receivers may be addressed using the same transmitter. It is available in 18 pin DIP and 20 pin SOP. The encoder will be in the Standby mode when the transmission is disabled. The various specifications of encoder HT12E are given in Table 3.7.

Table 3.7 Encoder HT12E Specifications

ENCODER HT12E SPECIFICATIONS	
Operating Voltage	2.4-12V
Operating Current	40-300 µA
Standby Current	0.1-4 µA
Output Drive Current	-1 to 1.6mA
Oscillator Frequency	3kHz

3.10 DECODER

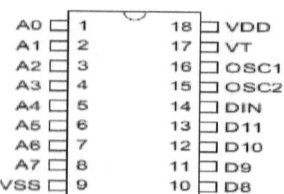

Fig 3.13 Pin Diagram of Decoder

The decoders receive serial addresses and data from a programmed 2^{12} series of encoders that are transmitted by a carrier using an RF or an IR transmission medium. They compare the serial input data three times continuously with their local addresses. If no error or unmatched codes are found, the input data codes are decoded and then transferred to the output pins. The VT pin also goes high to indicate a valid transmission. The details of various pins of decoder HT12D is shown in Fig 3.8.

The 2^{12} series of decoders are capable of decoding information that consist of N bits of address and 12-N bits of data. Of this series, the HT12D is arranged to provide 8 address bits and 4 data bits, and HT12F is used to decode 12 bits of address information. The 2^{12} series of decoders provides various combinations of addresses and data pins in different packages so as to pair with the 2^{12} series of encoders. The decoders receive data that are transmitted by an encoder and interpret the first N bits of code period as addresses and the last 12-N bits as data, where N is the address code number. A signal on the DIN pin activates the oscillator which in turn decodes the incoming address and data. The decoders will then check the received address three times continuously. If the received address codes all match the contents of the decoders local address, the 12-N bits of data are decoded to activate the output pins and the VT pin is set high to indicate a valid transmission. This will last unless the address code is incorrect or no signal is received. The output of the VT pin is high only when the transmission is valid. Otherwise it is always low. The various specifications of encoder HT12E are given in Table 3.7.

Table 3.8 Decoder HT12D Specifications

DECODER HT12D SPECIFICATION	
Operating Voltage	2.4-12V
Standby Current	1-4μA
Operating Current	400μA
VT output Source Current	-1 mA
VT output Sink Current	1mA
Oscillator Frequency	150kHz

3.11 POWER SUPPLY

The Arduino Uno can be powered via the USB connection or with an external power supply. The power source is selected automatically. External (non-USB) power can come either from an AC-to-DC adapter or battery. The adapter can be connected by plugging a 2.1mm center-positive plug into the board's power jack. Leads from a battery can be inserted in the Gnd and Vin pin headers of the POWER connector. The board can operate on an external supply of 6 to 20 volts. If supplied with less than 7V, however, the 5V pin may supply less than five volts and the board may be unstable. If using more than 12V, the voltage regulator may overheat and damage the board. The recommended range is 7 to 12 volts.

- VIN: The input voltage to the Arduino board when it's using an external power source (as opposed to 5 volts from the USB connection or other regulated power source). We can supply voltage through this pin, or, if supplying voltage via the power jack, access it through this pin.
- 5V: This pin outputs a regulated 5V from the regulator on the board. The board can be supplied with power either from the DC power jack (7 - 12V), the USB connector (5V), or the VIN pin of the board (7-12V). Supplying voltage via the 5V or 3.3V pins bypasses the regulator, and can damages the board.
- 3V: A 3.3 volt supply generated by the on-board regulator. Maximum current drawn is 50 mA.
- GND: Ground pins.
- IOREF: This pin on the Arduino board provides the voltage reference with which the microcontroller operates. A properly configured shield can read the IOREF pin voltage and select the appropriate power source or enable voltage translators on the outputs for working with the 5V or 3.3V.

An off-the shelf Arduino adapter can be used with the following features

- must be a DC adapter (between 9V and 12V DC).
- must be rated a minimum of 250mA current output, although 500mA or 1A output gives the current necessary to make each component of the circuit function reliably.
- must have a 2.1mm power plug on the Arduino end, and the plug must be "centre positive", that is, the middle pin of the plug has to be the + connection.

CHAPTER 4

SOFTWARE DESCRIPTION

The Arduino development environment contains a text editor for writing code, a message area, a text console, a toolbar with buttons for common functions, and a series of menus. This IDE connects to the Arduino hardware to upload programs and communicate with them.

4.1 WRITING SKETCHES

Software written using Arduino is called sketch. These sketches are written in the text editor. Sketches are saved with the file extension .ino. It has features for cutting/pasting and for searching/replacing text. The message area gives feedback while saving and exporting and also displays errors. The console displays text output by the Arduino environment including complete error messages and other information. The bottom right hand corner of the window displays the current board and serial port. The toolbar buttons allow you to verify and upload programs, create, open, save sketches, and open the serial monitor. Versions of the IDE prior to 1.0 saved sketches with the extension .pde. It is possible to open these files with version 1.0, save the sketch with the .ino extension on save.

4.2 ARDUINO IDE COMMANDS

The following commands are used in arduino IDE
- Verify: Checks your code for errors.
- Upload: Compiles your code and uploads it to the Arduino I/O board.
- New: Creates a new sketch.

- Open: Presents a menu of all the sketches in your sketchbook.
- Save: Saves your sketch.
- Serial Monitor: Opens the serial monitor to observe the output.
- Edit: Copy for Forum –Copies the code of your sketch to the clipboard in a form suitable for posting to the forum, complete with syntax colouring. Copy as HTML - Copies the code of your sketch to the clipboard as HTML, suitable for embedding in web pages.
- Sketch: Verify/Compile-Checks your sketch for errors. Show Sketch Folder- Opens the current sketch folder. Add File-Adds a source file to the sketch (it will be copied from its current location). The new file appears in a new tab in the sketch window. Files can be removed from the sketch using the tab menu. Import Library-Adds a library to your sketch by inserting #include statements at the start of the code.
- Tools: Auto Format- To format the user's code nicely. Archive/Sketch- Archives a copy of the current sketch in .zip format. The archive is placed in the same directory as the sketch. Board- To select the board that is being used. Serial Port- This menu contains all the serial devices (real or virtual) on your machine. It should automatically refresh every time you open the top-level tools menu. Programmer- For selecting a hardware programmer when programming a board or chip and not using the onboard USB-serial connection. Burn Bootloader- The items in this menu allow to burn a bootloader onto the microcontroller on an Arduino board. This is not required for normal use of an Arduino board but is useful when we use new Atmega microcontrollers which normally come without a bootloader. Ensure that you've selected the correct board from the Boards menu before burning the bootloader.

- Sketch: Verify/Compile-Checks your sketch for errors. Show Sketch Folder-Opens the current sketch folder. Add File-Adds a source file to the sketch (it will be copied from its current location). The new file appears in a new tab in the sketch window. Files can be removed from the sketch using the tab menu. Import Library-Adds a library to your sketch by inserting #include statements at the start of your code.

4.3 SKETCHBOOK

The Arduino environment uses the sketchbook: a standard place to store programs. The sketches in users sketchbook is opened from the File > Sketchbook menu or from the Open button on the toolbar. On running the Arduino software first time, it will automatically create a directory for user's sketchbook. Its location can be viewed or changed of the sketchbook location from with the Preferences dialog.

Fig 4.1 Arduino Sketchbook

Beginning with version 1.0, files are saved with a .ino file extension. Previous versions use the .pde extension. Users may still open .pde named files in version 1.0 and later, the software will automatically rename the extension to .ino. In Fig 4.1 an arduino sketchbook is shown. Tabs, Multiple Files and Compilation-Allows to manage sketches with more than one file (each of which appears in its own tab). These can be normal Arduino code files (no extension), C files (.c extension), C++ files (.cpp), or header files.

4.4 UPLOADING

Before uploading the sketch with code, one needs to select the correct items from the Tools > Board and Tools > Serial Port menus. Once the user selects the correct serial port and board, press the upload button in the toolbar or select the Upload item from the File menu. Current Arduino boards will reset automatically and begin the upload. With older boards that lack auto-reset, it need to press the reset button on the board just before starting the upload. On most boards, the RX and TX LEDs blink as the sketch is uploaded. The Arduino environment will display a message when the upload is complete, or show an error. When a user uploads a sketch, the Arduino bootloader being used, is a small program that has been loaded on to the microcontroller on the arduino board. It allows the code to be uploaded without using any additional hardware. The bootloader is active for a few seconds when the board resets; then it starts whichever sketch was most recently uploaded to the microcontroller. The bootloader will blink the on-board LED when it starts (i.e. when the board resets).

4.5 LIBRARIES

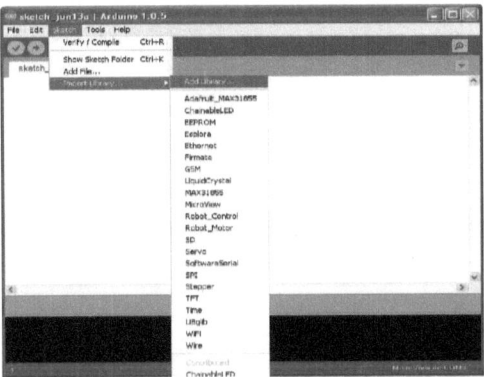

Fig 4.2 Importing a library into Arduino IDE

The Fig 4.2 shows how to add a library file for the current sketch. Libraries provide extra functionality for use in sketches, e.g. working with hardware or manipulating data. To use a library in a sketch, select : Sketch > Import Library menu. This will insert one or more #include statements at the top of the sketch and compile the library with your sketch. Because libraries are uploaded to the board with your sketch, they increase the amount of space it takes up. If a sketch no longer needs a library, simply delete its #include statements from the top of the code. Some libraries are included with the Arduino software. Others can be downloaded from a variety of sources. Starting with version 1.0.5 of the IDE, user can import a library from a zip file and use it in an open sketch. Support for third-party hardware can be added to the hardware directory of your sketchbook directory. Platforms installed there may include board definitions, core libraries, bootloaders, and programmer definitions. To install, create the hardware directory, then unzip the third-party platform into its own sub-directory. To uninstall, simply delete its directory.

4.6 SERIAL MONITOR

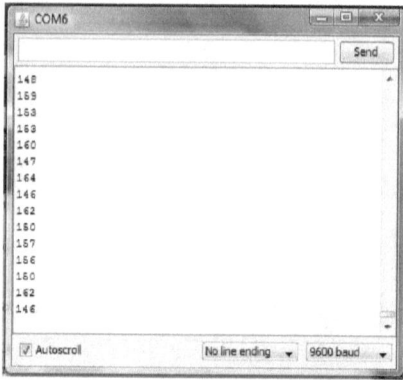

Fig 4.3 Serial Monitor

Serial monitor displays the serial data being sent from the Arduino board (USB or serial board) as shown in Fig 4.3. To send data to the board, enter text and click on the "send" button or press enter. Choose the baud rate from the drop-down that matches the rate passed to that of the "Serial.begin" in the sketch. In Mac or Linux, the Arduino board will reset (rerun your sketch from the beginning) when you connect with the serial monitor.

CHAPTER 5

SYSTEM IMPLEMENTATION AND RESULTS

5.1 TRANSMITTER SECTION

The MPU-6050 gyroscope module is a low power 6 DOF sensor with temperature data. To test the functioning of the MPU-6050 gyroscope the connection used is as follows GND-to be connected to Arduino's GND, VCC-to be connected to Arduino's 5V, SDA and SCL to A4 and A5 respectively.

If each axis is placed in the same plane the values should be the same. By varying the sensitivity the analog values will be different. With the axis horizontal (i.e. parallel to ground or 0°), the reading from this sensor should be around 400, but values at other angles will be different. The range is from 0 to 65535 as MPU-6050 has 16bit ADC. Fig 5.1 shows typical gyro sensor values displayed in serial monitor.

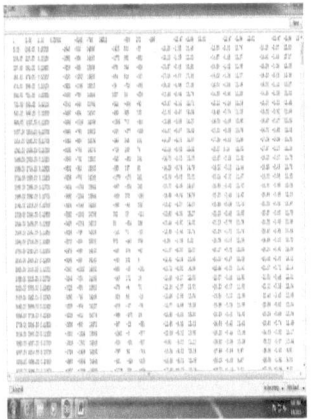

Fig 5.1 Output of MPU-6050

The KY-308 microphone sound sensor is a very affordable distance sensor mainly used for object avoidance. It essentially gives the Arduino about special awareness and can detect falling of objects/persons. To test the working of this sensor the connection used is as follows GND-to be connected to Arduino's GND, VCC-to be connected to Arduino's 5V and pin A0 to Arduino's A0(for analog program) and pin D0 to Arduino's 13(for digital program). The sensitivity of this sound sensor can be tuned for different applications. This sensor satisfies it's test condition when it detects sound for 5seconds or more.

The piezo vibration sensor is useful to find the impact of fall or shock. The pin connections to be used are + pin to any analog pin in arduino (here A0 is used) and – pin to GND in Arduino. The sensor values range from 0 to 1023 and the values beyond 800 is set as threshold level for this sensor. The output of piezo vibration sensor is given in Fig 5.2.

The connection used for RF transmitter is as follows: GND-to be connected to Arduino's GND, VCC-to be connected to Arduino's 5V, the data pin to any digital pins in Arduino board. But when used along with encoder a separate 4*1 connection to the digital pins (barring digital pins 0 and 1) is made.

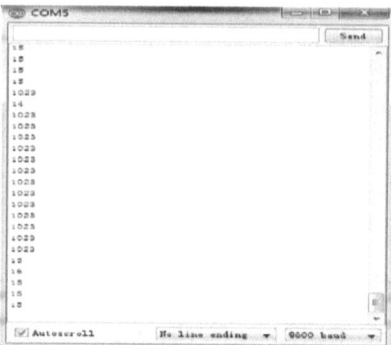

Fig 5.2 Output of Piezo vibration sensor

Here the transmitter section consists of gyro sensor, piezo vibration sensor, microphone sound sensor an Arduino Uno board and a RF 433 MHz transmitter with an encoder. The entire transmitter section can be represented by a block diagram as in Fig 5.3.

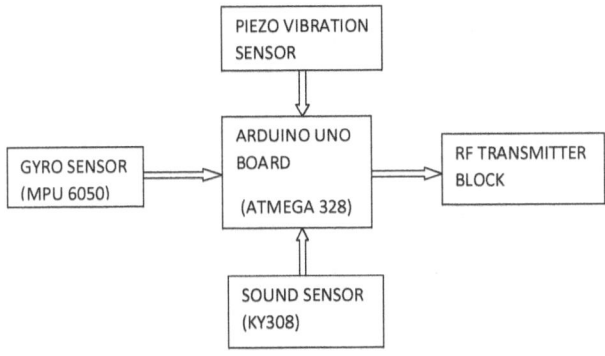

Fig 5.3 Block Diagram of Transmitter Section

5.2 RECEIVER SECTION

The receiving section as shown in Fig 5.4 consists of a microcontroller board to read the RF receiver values and a GSM module to send alert to the care-giver upon reaching the threshold values set for emergency alert. The Arduino board's microcontroller is used to check and execute the test conditions.

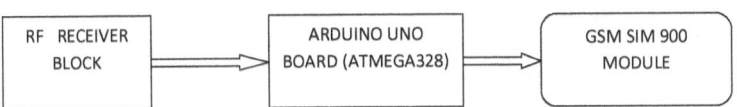

Fig 5.4 Block Diagram of Receiver Section

The functioning of GSM module is checked by making a call to the SIM in it. GSM module works when the ring goes to this module as in mobile phone. There are two LEDs in GSM module to indicate the presence of power and signal coverage when power connections are made to it.

5.3 FLOWCHART OF THE SYSTEM

The flowchart in Fig 5.5 gives an outline about the working of the entire system.

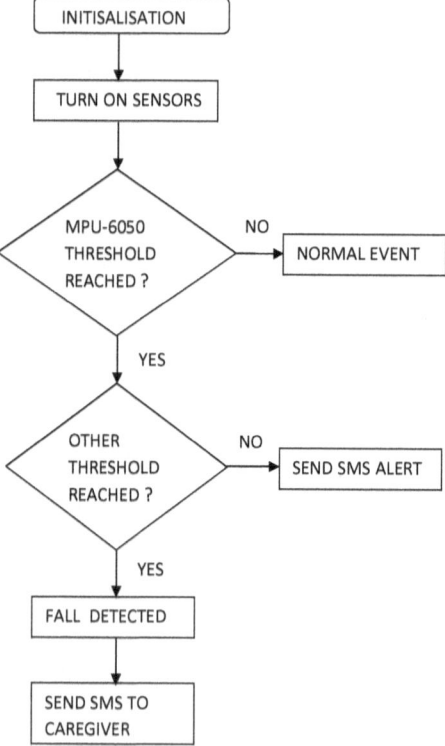

Fig 5.5 Flowchart of the System

5.4. RESULTS AND DISCUSSIONS

The technique of detecting fall relies on deriving human body posture with a suitable sensor placed at appropriate anatomical position and an effective algorithm which precisely distinguish daily activities and fall. Since, change of plane of body and sudden changes in acceleration and orientation are involved in fall; MEMS gyro sensor is a suitable choice. The changes of acceleration in 3 axes and orientation in another 3 axes are monitored continuously. When the changes in acceleration fall in the window of threshold, the event is checked for a fall activity.

The components are enclosed in a plastic box and tied in the waist region of the volunteers. Each subject is made to fall on bed, and many fall like activities (picking objects from the floor and walking on staircase). The experiment is performed on 5 adult male subjects (age from 19 to 25 years, weight from 50 to 80 kg, and height from 154.5 - 180.0 cm). In this experiment, subject performs ADL such as standing, walking, sitting down/ standing up, step, run, and 4 different kinds of fall tests: forward fall, backward fall, right sideway fall, and left sideway fall.

Most falls occurred on a bed, which suggests that the impact could potentially differ from falls on harder or unexpected surfaces. Even with the assumption of a vertically arranged z-axis the results for the directional fall identification are satisfactory. However, this assumption was made under the restrictions of sensor positioning.

Out of fifty trials (ten trials each subject) taken for the analysis the events are classified as TP (True Positive), TN (True Negative), FP (False Positive), FN (False Negative). The trials included forward fall, falling sideways and backward fall. The missed subjects knelt down slowly and leaned forward without giving any jerk. Since most of the falls not likely to happen this way, the missed detections do not have significance.

The fall simulating activities like jogging, skipping, and picking objects did not create any false alarms. The outcome of the experiments is listed in the table 5.1. As we see from the table the accuracy is 86%.

Table 5.1 Analysis of prototype testing

Volunteer	Number of trials	Number of fall detections	Number of false alarms
1	10	9	1
2	10	9	1
3	10	8	2
4	10	8	2
5	10	9	1

The sensitive and the specificity of the system are defined as follows:

$$\text{Sensitivity} = (\text{No.TP})/(\text{No.TP} + \text{No. FN}) \qquad (5.1)$$

$$\text{Specificity} = (\text{No. TN})/(\text{No. TN} + \text{No. FP}) \qquad (5.2)$$

☐ Number of True positive (No.TP): a fall occurs, the device detects it.

☐ Number of False positive (No. FP): the device announces a fall, but it did not occur.

☐ Number of True negative (No.TN): a normal (no fall) movement is performed, the device does not declare a fall.

☐ Number of False negative (No. FN): a fall occurs but the device does not detect it.

Table 5.2 Accuracy, Specificity, Sensitivity of the falls

Sensors Used	Accuracy	Specificity	Sensitivity
Accelerometer + Ultrasonic sound sensor	81	83	79
Gyro sensor+ Piezo vibration sensor + Microphone sound sensor	86	85	87

All the modules are powered using 12V DC adapter. The programs to test sensors, other system modules and final system implementation to test conditions are written in embedded C. The code is uploaded into the Arduino board using the USB cable. Reset the Arduino boards upon powering. Each sensor used in this project is used to give a test condition based on which SMS alert may be sent to the care-giver. The threshold values are selected based on readings of sensor for different tilted positions or impact of vibration or sound.

Table 5.3 Threshold levels for sensors

S.NO	SENSOR	THRESHOLD VALUE
1	Gyro sensor	Beyond 10000 readings (for X and Y axes)
2	Piezo Vibration sensor	Beyond 800 readings
3	Microphone sound sensor	Sound or noise detected for 5 seconds or more

The results for these test conditions are observed in serial monitor of the transmitter side Arduino board. Once the test condition for gyro sensor alone reached we will observe an output "Fall on". For this purpose a flag condition "Fall on =1" must be achieved. When the Arduino board recognizes the flag condition "Fall on Emg=1" then it will display the output as Fall on Emg. This flag condition occurs when threshold level for gyro sensor is satisfied along with threshold level for piezo vibration sensor or sound sensor or both of these sensors.

As well as once threshold level of gyro sensors reached, we will also observe the serial monitor output as Front /Back (for X-axis) or Right/ Left (for Y-axis) depending upon the direction of inclination of gyro sensor with respect to the reference plane. A serial monitor output for these flag conditions are shown in fig 5.6.

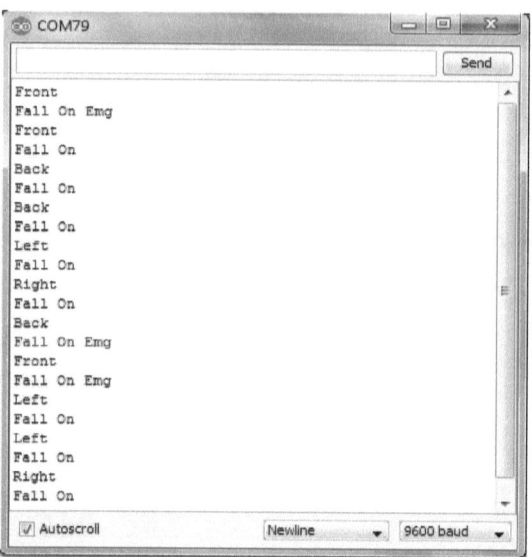

Fig 5.6 Output for test condition

Based on the test conditions reached in the Arduino board, the GSM module sends the text message "Fall on" or "Fall on Emg". The specificity and accuracy of the system depends upon the position in which it is placed on the body. As we used gyro sensor which provides reliable orientation of an object and also this system depends on the tilting, the most viable location to place the transmitter side module is waist. Also from the previous researches we observe that waist region is suitable to place this module as it identifies posture for sitting/standing /bending or lying.

5.5 FALL DETECTION IN ADL

In addition to detecting the occurrence of falls, it is also essential to find whether the system identifies false alarms from various day to day activities. This analysis will help to improve the efficiency and specificity of the system by modifying the test conditions to avoid false alarms. Table 5.4 lists few ADLs.

Table 5.4 Fall detection

S.NO	ACTIVITY	FALL DETECTED
1	Sitting down on a chair	No
2	Standing up from the chair	No
3	Resting against a wall, then sliding down	No
4	Jumping	No
5	Searching something on the floor	No
6	Bend forward	No
7	Walking on the floor	No
8	Walking on the stairs	No
9	Forward collapse(Fall on knees)	Yes
10	Forward collapse (Lie down)	Yes
11	Backward collapse	Yes
12	Sidewise collapse	Yes
13	Fall from the chair	Yes
14	Fall from bed	Yes
15	Collapsing into a bed	Yes
16	Falling Slowly	Yes

The different types of fall may vary in direction, impact of fall, tilt position and angle, anatomical position in which the module is kept. Accordingly sensitivity and specificity of the system will vary.

CHAPTER 6

CONCLUSION AND FUTURE WORK

6.1 CONCLUSION

The proposed project gives an easy mechanism to detect falls in elderly persons using user friendly tools. Our solution does not need complex computation, so the detection process can be implemented simply. This provides an opportunity for the immediate medical attention to the elderly at the earliest once the fall detection alert is sent. The performance under real-life conditions, usability and user acceptance as well as issues related to power consumption, real-time operations, sensing limitations, record of real-life falls are analysed using this kit.

This work effectively utilizes gyroscopes and accelerometer-derived posture information to detect fall events. It also features low computational cost and fast response. However, this method has difficulties in differentiating various postures as the vibration sensor and gyro sensor are more sensitive even to small changes. To distinguish these activities, context information (environmental/physiological) are needed.

6.2 FUTURE ENHANCEMENTS

The system can be implemented with inclusion of advanced 9DOF gyro sensor to improve its efficiency. Gyroscope, along with accelerometer, performs better than accelerometer alone. Blood pressure sensor and heart beat sensor provides the state of the person during normal and fall events. Mobile apps and Internet of things when employed for such systems will make it more interactive.

APPENDIX

TRANSMITTER SECTION

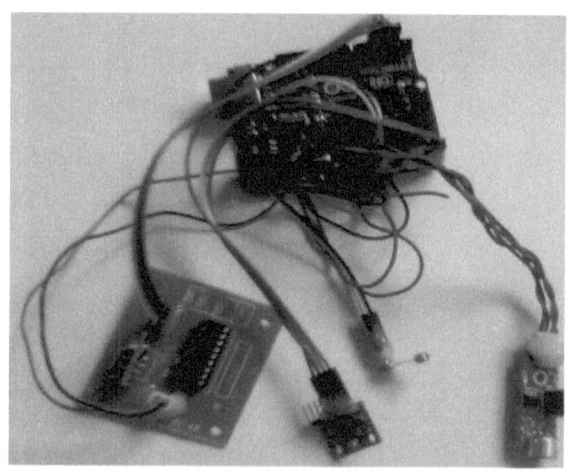

RECEIVER SECTION

REFERENCES

[1] Aldo Moro, Eric Mercier, Gilles Morey, Gilles Virone, Norbert Noury, Thieny Porcheron, Thierry Herd, Vincent Rialle (2000) 'Monitoring Behavior in Home Using a Smart Fall Sensor and Position Sensors' – 1st Annual International IEEE-EMBS Special Topic Conference on Microtechnologies in Medicine & Biology,Vol.5 No.6, pp.607-610.

[2] Brian Ahearne, Cliodhna Ni Scanaill, Gerard M. Lyons (2006) 'Long –Term Telemonitoring of Mobility of Elderly People Using SMS Messaging'-IEEE Transactions on Information Technology in Biomedicine Vol.10, No.2, pp.412-413.

[3] Jing Hou, Jue Wang, Ping Liu and Tong Zhang (2006) 'Fall detection by Embedding an Accelerometer in Cellphone and Using KFD Algorithm' - International Journal of Computer Science and Network Security, Vol. 6, No. 10, pp. 277-284.

[4] A.K. Bourke, A. Fleury, G. ÓLaighin, N. Noury, P. Rumeau, V. Rialle, J.E. Lundy (2007) 'Fall detection – Principles and Methods' - Proceedings of the 29th Annual International Conference of the IEEE EMBS Vol.4, No.3, pp. 1663-1666.

[5] Hoda Eydgahi, Ming-Zher Poh and Richard R. Fletcher (2010) 'Wearable Sensors: Opportunities and Challenges for Low-Cost HealthCare' - 32nd Annual International Conference of the IEEE EMBS, pp. 1763-1766.

[6] Bin Li, Jin Wang, R. Simon Sherratt, Sungyoung Lee, Zhongqi Zhang (2014) 'An Enhanced Fall Detection System for Elderly Person Monitoring using Consumer Home Networks' – IEEE Transactions on Consumer Electronics Vol.60, No.1, pp.23-29.

[7] Afshin Nabili, Binh Q. Tran, Su V. Tran, Uyen D. Nguyen, Quoc T. Huynh (2013) ' Fall Detection system Using Combination Accelerometer and Gyroscope'- Proceedings of the second international conference of advances in Electronic Devices and Circuits, pp.52-56.

[8] www.analog.com

[9] www.atmel.com

[10] www.ti.com

[11] www.invensense.com

[12] www.smssolutions.net

[13] www.sim.com

[14] www.arduino.cc

[15] www.meas-spec.com

[16] www.starlino.com

[17] www.varesano.net

[18] www.i2cdevlib.com

I want morebooks!

Buy your books fast and straightforward online - at one of the world's fastest growing online book stores! Environmentally sound due to Print-on-Demand technologies.

Buy your books online at
www.get-morebooks.com

Kaufen Sie Ihre Bücher schnell und unkompliziert online – auf einer der am schnellsten wachsenden Buchhandelsplattformen weltweit! Dank Print-On-Demand umwelt- und ressourcenschonend produziert.

Bücher schneller online kaufen
www.morebooks.de

OmniScriptum Marketing DEU GmbH
Heinrich-Böcking-Str. 6-8
D - 66121 Saarbrücken
Telefax: +49 681 93 81 567-9

info@omniscriptum.com
www.omniscriptum.com